Number and Operations

Homes of All Shapes

Homes are different shapes and sizes.
Homes give shelter.

What shape is your home?

Some homes look like triangles.
These houses are A-frame houses.

A-frame houses are good where there is a lot of snow. The shape of the roof lets the snow slide off.

One-floor houses also have shapes. They can be shaped like a long rectangle. They can be shaped like the letter *T*.
Can you think of a home that is one of these shapes?

Some two-floor houses are shaped like a square. These houses have an upstairs and a downstairs.

Some people live in townhouses or tall buildings. These homes are sometimes rectangular.

What shapes do you see in this building?

Did You Know?

Some Native American tribes lived in teepees. These homes were shaped like cones. Others were rectangular, square, or even round.

What shapes are these homes?

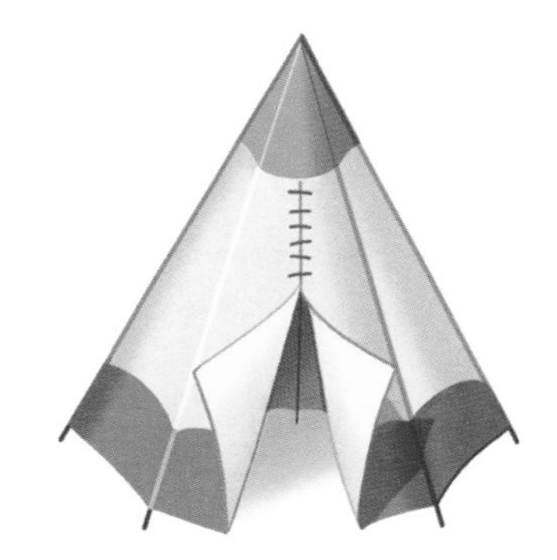

People who camp might sleep in a trailer. They might sleep in a tent.

Look at the pictures. Can you find a triangular shape? Can you find a rectangular solid?

Animal homes also are different shapes and sizes.
Look at the homes on this page.

A gopher lives under the ground.

Many other animals dig tunnels like this.
It is a cylinder-shaped tunnel.

Bald eagles make big nests.
The bottom of the nest is
shaped like a triangle.

Robin's nest

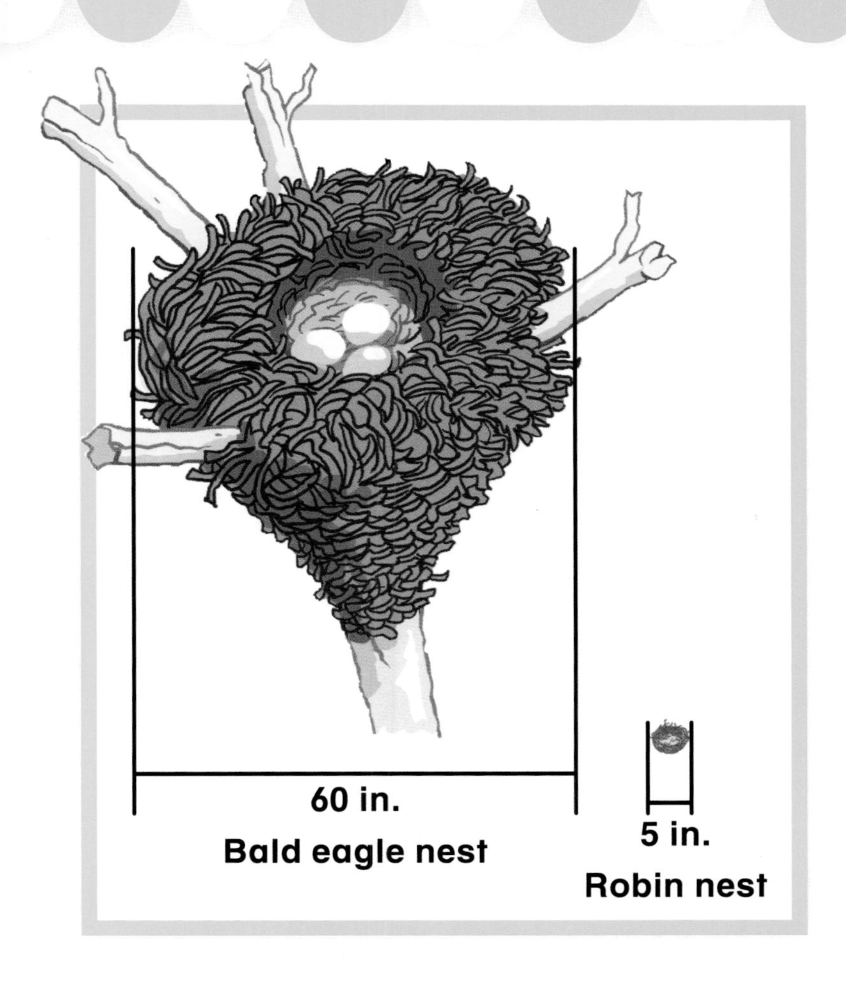

The eagle's nest is messy but strong. It can be 2 feet deep and 5 feet across! Look how big it is.
A robin's nest is less than 5 inches across.

Have you ever seen a beehive?
Honeybees live in beehives.
Their homes are made up of many parts. Each part is shaped like a hexagon.

Sometimes people and animals live in the same home. Sometimes there are more pets than people!

Some dogs live in a house outside. What shape might it be?

Fish, birds, and hamsters are also pets.
They have little homes.

Some animals carry their home with them.

What other animal homes can you think of?
What shapes are they?

Animals and people have one big home.

Our home is Earth!

JIM BRANDENBURG

To the Top of the World

ADVENTURES WITH ARCTIC WOLVES

Edited by JOANN BREN GUERNSEY

WALKER AND COMPANY NEW YORK

First published in the United States of America in 1993 by Walker Publishing Company, Inc.; first paperback edition published in 1995.

Published simultaneously in Canada by Thomas Allen & Son Canada, Limited, Markham, Ontario

Library of Congress Cataloging-in-Publication Data
Brandenburg, Jim.
To the top of the world: adventures with Arctic wolves / Jim Brandenburg: edited by JoAnn Bren Guernsey.
p. cm.
Summary: A wildlife photographer records in text and photographs a visit to Ellesmere Island, Northwest Territories, where he filmed a pack of Arctic wolves over several months.
ISBN 0-8027-8219-1. —ISBN 0-8027-8220-5 (lib. ed.)
1. Wolves—Northwest Territories—Ellesmere Island—Juvenile literature. [1. Wolves.] I. Guernsey, JoAnn Bren. II. Title.
QL737.C22B636 1993
599.74′442—dc20 93-12105
CIP
AC

ISBN 0-8027-7462-8 (paper)

Book design by Victoria Hartman

Printed in Hong Kong

10 9 8 7 6 5 4 3 2 1